On Cue

On Cue

Cristy Watson

Orca currents

ORCA BOOK PUBLISHERS

Library and Archives Canada Cataloguing in Publication

Watson, Cristy, 1964-, author
On cue / Cristy Watson.
(Orca currents)

Issued in print and electronic formats.
ISBN 978-1-4598-1105-8 (paperback).—ISBN 978-1-4598-1106-5 (pdf).—
ISBN 978-1-4598-1107-2 (epub)

I. Title. II. Series: Orca currents
PS8645.A8625052 2016   jc813'.6   C2015-904504-5
C2015-904505-3

First published in the United States, 2015
Library of Congress Control Number: 2015946244

Summary: Fourteen-year-old Randi has to balance theater
studies with caring for her brother with autism.

*Orca Book Publishers is dedicated to preserving the environment and has
printed this book on Forest Stewardship Council® certified paper.*

Orca Book Publishers gratefully acknowledges the support for its
publishing programs provided by the following agencies: the Government
of Canada through the Canada Book Fund and the Canada Council for the Arts,
and the Province of British Columbia through the BC Arts Council
and the Book Publishing Tax Credit.

Cover photography by Dreamstime.com
Author photo by Lynne Woodley

ORCA BOOK PUBLISHERS
www.orcabook.com

Printed and bound in Canada.

19 18 17 16 • 4 3 2 1

This book is dedicated to my mom and dad. Thanks for your help in making this a better story and for your encouragement along the way. You've helped nurture me as a writer.

In memory of Uncle Bob, 1932–2015

Chapter One

I undo the braid in my hair and work my fingers through the auburn waves. As loose hairs fall to the floor, Mom gives me a look that says, *Not at the breakfast table*. My shoulders slump as she lays a bowl of soupy oatmeal in front of me. My younger brother, Toby, is loading his spoon with only the pink Froot Loops. He has the morning paper in front of him.

"*Matthews, Karen. Died August 22. She is survived by her brother…*" he reads.

"Does he have to do this every morning?" I ask.

"*He* has a name." Mom dips a piece of toast in her oatmeal. Some spills onto her skirt. "Damn, I don't have time for this."

"Damn," repeats Toby. "*Peters, Shirley. Died…*"

"Quit it, Toby."

"*Tobias*," he says and jams another spoonful of the pink cereal into his mouth.

My sigh goes unnoticed. After one reading, Toby will have the obituaries memorized. Then he'll repeat them all day long.

"Randi, I'm going to be late. If you and Toby don't hurry, you'll be late too. Is that how you want to begin the school year?"

The last thing I want is to be late for my first day of high school. As I swallow the gray goop, Mom finishes her toast. Before she heads out of the kitchen, she gives Toby a slurpy kiss on the top of his head. He smooths his oily black hair back into place.

I *was* looking forward to grade eight. I would finally have freedom. Finally get away from the responsibility of looking after my brother all day long. I hear what the other kids say when we pass by. *There goes that girl and her brother. Did you hear him wailing in the assembly last year? Do you know he'll repeat swearwords if you say them?* Then they spew a bunch of bad words and wait for Toby to repeat them. Laughter usually follows.

They judge me by my brother. High school was going to be my chance to stand on my own.

Then Mom crashed my party. I have to walk Toby to and from his school. Every day. That means five blocks out of the way. That means the end of my social life.

No chance to be normal.

I hear the door slam as Mom leaves for work. "Come on, Toby," I say. "Finish your breakfast so we aren't late." I still have to fiddle with my new contact lenses.

"*To...bias*," he replies, then gets up and puts his bowl on the counter instead of in the dishwasher. I'm about to reprimand him when I notice his hands flap. He begins to rock back and forth on the balls of his feet. Giving him heck when he's in this state might put him over the edge. And then I'll definitely be late.

I take a deep breath so I don't sound mad. "Remember, Ms. Banyan is your teacher again this year. And your favorite staff, Miss Maureen, will be waiting for you. Just like always."

"Maureen loves turtles," he says. I help him tie his shoes. He stops rocking but still flaps his hands. He pats me on the head as I finish. "You will be in grade eight."

"That's right," I answer. "Remember, you have to wait for me after school." I doubt he will forget. Toby has waited for me every day for three years. This year I have to leave early from my last class to get to Toby's school for the bell. He works best with solid routines.

I brush my teeth but don't stress about getting Toby to do his. I don't need the hassle.

Putting my contacts in is tricky. Not only is this the third time I have *ever* put them in, but my hands are sweating as I think about school. I don't want to lose a lens down the drain. I asked the doctor a million times if the contacts can slip *behind* my eyes. He said no, but I place them on each of my green eyes slowly, just to be sure.

The first day of high school would be easier if I still had my best friend.

But Laurel moved to Calgary over the summer. As I put my dangly earrings in, I focus on the one good thing about being in high school—we have an elective. I chose drama. I finally get to pursue my dream of being an actress.

Mr. Dean will be our drama teacher. I met him at orientation, and he is *super* cute.

By the time I return to the kitchen with my knapsack, Toby has untied one shoe.

"We don't have time for this crap," I say.

"Crap. We don't have time."

"Come on, Toby. *Don't do this*." I hear my mother's voice reminding me to be patient with him. "We went to school every day last year. I drop you at your class. Then I pick you up at the end of the day." I slide my feet into my new flats and tie his shoe again.

The sun hits us as we head out the door. September is usually a hot month in Vancouver. Now I wish we had three months for summer vacation.

Toby has pulled his knapsack off by the time we reach his school. He claims it's too itchy. Before we enter his class, he tugs on my shirt. Miss Maureen comes to the door and greets us.

"Welcome back, Tobias. Great to see you. Hope you had a good summer?" She takes his knapsack from me and puts one hand on his shoulder. He grabs my hand and squeezes it tightly. He doesn't like to be touched.

"You know everyone here," I whisper. "You know Miss Maureen, and there is Ms. Banyan, see?" I point to his teacher. Toby is in a split class, so that he can have consistency with the same teacher and support worker he had last year. They've both been awesome with him.

Ms. Banyan waves our way. I gently guide Toby into the room.

"Tobias, I am happy you're here. Missy will help you with your things. You'll share a space in the cloakroom." A girl I remember from last year skips our way. Toby lets my hand go and follows Missy into the cloakroom.

"See you after school," I say to his back. Miss Maureen waves, and I fly down the hall and out the door. I make it to my homeroom as the second bell rings. I am dripping with sweat, and I didn't have a chance to check my hair in the bathroom. Now I wish I had worn my hoodie—then I could hide. Note to self: leave home earlier to have time to fix hair and check clothes!

We have a TA system at our school, which means the homeroom is made up of students from grades eight to twelve. It turns out Mr. Dean is also our teacher-advisor. "Welcome back to the returning students and glad to have you aboard, grade eights," he says. "Just a few notes to share with you before we get ready for first block."

I shuffle uneasily in my seat. I can feel someone's eyes boring into me from three seats over, but I can't make out the person's face in my peripheral vision. I'm contemplating taking out my makeup mirror so I can see who is staring at me when Mr. Dean drops a bomb.

"The school board had to make some tough decisions, and one of them was to cancel the drama program for grades eight and nine. But don't worry. You can take it in grade ten."

So much for having any fun in high school.

Chapter Two

With Mr. Dean's news, several students shout out their disappointment.

I feel deflated. Drama was the one thing I was looking forward to this year. As I wonder which class will replace it, a grade-twelve student passes me a sheet with my new elective. I am now in home economics. *Crap!* I have to learn to sew. Who sews? Your socks get a hole, you throw them out.

Even if I'd gotten an elective I like, it wouldn't take away the sting of losing drama.

"Believe me, I am just as disappointed," says Mr. Dean. "I lose two blocks of teaching a subject I love."

"Isn't there anything we can do? Talk to someone maybe? Like someone on the school board?"

I turn my head to see who owns the voice. It's the guy who was staring at me a few minutes ago. His dark eyes shoot a glance my way, and I look at the floor. I fiddle with the green streak in my hair and listen to the conversation.

"That's a great idea, Josh. But I don't think it will have any impact. The numbers aren't going to change between now and this afternoon, and the district won't give any more money to the school."

"Well, there are twenty of us in homeroom. That means twenty possible ideas. Let's brainstorm and see what we come up with. Are you game, Mr. Dean?"

I peek at Josh. He's smiling at the class, proud of his plan. He has perfect teeth and dimples. Instantly I like his style—a white shirt under a grey sweater, with a skinny black tie loosely knotted around his neck.

I scan the room and realize I don't know any of the students. Four elementary schools feed into our high school. A girl wearing a hijab says we should at least be allowed to have our second or third choice for the replacement. I agree. I had art and psychology for my other choices. Mr. Dean takes a vote, and everyone in the class supports us.

Josh puts up his hand.

"Mr. Dean, I know that after school is a time when you probably do marking or stuff like that, but could drama be offered as an extracurricular activity? Maybe even for…extra marks?"

Several students whistle and clap. I smile. What a great idea!

"I know for certain that no extra marks will be given to any course that happens after school," says Mr. Dean. "None of the sports teams receive extra credit. But I do think you have something there."

"You mean you'll consider teaching it?" asks Josh.

"Well, I'm here until five PM anyway, so I am available…Sure, why not? Let me run it past our admin team. If they are okay with it, so am I."

The class erupts in cheers. Mr. Dean smiles. I jump out of my seat and watch Josh, who I decide must love acting, or he wouldn't have taken up this cause. He's smiling my way.

The bell rings for our first class, and I don't have a locker yet. None of us do.

We were supposed to be sorting out lockers instead of rallying to replace our drama class. Then it hits me. Not having a locker is the least of my worries.

We just asked for drama to be after school. My mom works until four thirty. I have to look after Toby.

How will I do both?

Chapter Three

In Math and English I sit with a girl named Amber. She just moved here this weekend from Burnaby.

"I haven't unpacked yet. Want to come help me after school?" she asks.

We're in our last block, and it is fifteen minutes before the bell. I have to leave now if I want to get to Toby on time. "That would be awesome, but I can't."

"What's up?" She rolls her tongue over her braces. I immediately feel an uncomfortable sensation roll down my shoulders, and I shudder. I'd be worried about cutting my tongue.

"I have to watch my brother after school," I answer.

"Oh, he's younger?"

"Yeah. Sort of…" I don't want to tell her he's in grade six. She'd wonder why he can't go home on his own.

"Listen," I continue, so she won't ask more questions. "Are you planning on attending drama if it happens after school?"

"You bet," she says. "I don't know if I'm any good at acting, but I hope there will be a small part for me. I like hanging out with actors. They're cool. Like you."

I blush. I've certainly never thought of myself as cool. I'm usually shy, but I've found in school plays that it's a lot easier when I play a role. I'm more confident when I'm being someone else. "I think I'll talk to my mom about finding someone to watch my brother after school tomorrow. Then we can fix up your room or hang out in drama, if it gets approved."

"It's a plan," she says and high-fives me.

The teacher asks if *we* would like to teach the class, since we seem to have something more interesting to share. I blush again, then apologize to Mrs. Wilkins as I pass her the note giving me permission to leave early.

"Off you go then," she says. "Make sure you finish your essay for tomorrow."

I rush down the hall. We got our lockers at lunch, and I've already forgotten my combination. I rustle through my papers but can't find it. Three tries later I still can't open it. Now I am flustered, because I can't be late to pick up Toby. He'll have a fit if I'm not there when the bell goes. I abandon the locker and carry all my books in my arms, my knapsack tucked away in the metal closet at school.

"Cheese and crackers," Toby chants all the way home. I'm trying to figure out what to do with him if we have drama after school. Asking Mom is not going to work. She's never trusted anyone with Toby before, so why would she change her mind now?

As we pass the MacGregors' house, I stop.

"Toby, I'll fix cheese and crackers for you and I'll let you play on my iPad if you stay in the house. I have to run back to the MacGregors' for a minute. Deal?"

"Deal," he says. He loves my iPad, especially the *Plants vs. Zombies* game. He's amazing at it too. He seems to know intuitively where to put each kind of plant so he can attack the zombies. He is on the gazillionth level, while I'm stuck on level eight. I've asked him to teach me, but he can't explain how he knows where to put the plants on the grid.

Once in the kitchen, I cut some mild cheddar cheese into triangles and put them on Triscuits. Toby will only eat them if the cheese is melted, so I place them in the microwave for thirty seconds. He's already advancing a level when I set the plate in front of him.

"I'll be right back," I assure him. He laughs as he wins tons of coins.

I have no problem convincing Sanya to watch Toby tomorrow after school. I figure even if we don't have drama, it will be cool to hang out with Amber. Sanya knows Toby has special needs. She agrees to watch him once I say Toby can borrow my iPad, which means he probably won't even know she's there, *and* I will pay her twenty dollars.

I just have to make sure Mom doesn't find out.

Now I have to cross my fingers that Mr. Dean gets permission to have drama after school. My new friend, Amber, will be there. And probably so will Josh.

Grade eight is shaping up to be a fabulous year!

Chapter Four

"So is there any news, Mr. Dean?" Josh asks as soon as we take our seats for homeroom.

"Yes. There is news..." Mr. Dean hesitates. He looks at the floor and shakes his head.

I feel tears well up in my eyes. I was sure we would have permission. It's not like it's costing the school any money.

I'm ready to rally when the class begins to laugh. I look up at Mr. Dean and realize he's teasing us.

"Sorry about that. I enjoy a chance to rehearse my skills. Guess I still have it." He smiles. "Okay, any students interested in taking drama are to meet in the theater after school today. Oh, and as you probably guessed, we will combine grades eight and nine."

Josh is grinning from ear to ear, and Amber smiles my way. I'm so glad I made arrangements for Toby after school. He was upset at first. But I gave him my iPad, and he settled down.

"Are you sure your mom is okay with Sanya picking Toby up?" asked Miss Maureen when I dropped him off. I hadn't thought about the school asking questions, so I was thrown off guard.

"Ah...I didn't have time to check with her. But Miss Maureen, this is

really important. It's about a class I have at school."

"And they expect you to stay late?"

"Yes."

I hope that doesn't count as lying. But becoming an actress is the most important thing in my life, so I had to be ready in case the school approved of our plan.

My answer seemed to satisfy her, and Toby skipped into class. They have a pet this year, a small white rabbit named Hopscotch. Toby loves holding him.

With my brother covered, I allow myself to be excited. In fact, I realize my nerves are heating up.

Now that drama is a reality, I worry about how I will do as an actress. Will I be any good? What play will we do? What role will I get? What will Josh think of me? Being in the same room as he is causes a full gymnastics routine in my stomach.

The day takes forever to end. When the final bell rings, I run to the drama room. Several grade-nine students are already there, including Josh.

"Hey, Randi," he says as he holds out his hand to help me onto the stage. His hand is warm. I notice the stairs off to the side after I thank him.

Mr. Dean is several minutes late. He apologizes, then breaks us up into groups to consider options for which play to perform. Each group receives a list of plays to choose from.

I am in a group with Isabella. I remember her from grade five, but she hasn't been in my class for the last two years. She isn't my favorite person in the world. But maybe she is different now that we're in high school.

"We better get started so we have a say," says Isabella as our group gathers on the left of the stage. "And *you're* taking drama?" she asks me.

I guess she isn't any different.

Isabella controls most of the discussion. She scoffs at all our suggestions. I agree with several students. The list is kind of lame. I don't recognize many of the titles. I wish they had a fantasy or Fey play. I love movies and TV shows with vampire love stories.

"Well, let's hear ideas from each person," says a grade-nine student I don't know. We haven't introduced ourselves. Isabella steals a look my way that makes me uncomfortable.

As we go around the group with suggestions, I find it tough to concentrate. I wonder how Sanya is doing with Toby. Will my plan work? I don't like to talk in front of strangers. I dread having to speak when I can't keep my thoughts straight. That's usually when I say something stupid. As it comes to my turn, my buzzing cell phone saves me.

"You shouldn't have your phone on during class," says Isabella with a smirk. "It's rude."

I ignore her and turn away from the group to read the text. It's from Sanya. All it says is, **Get home NOW!**

Chapter Five

"Sorry, dudes, I have to bail."

"What about your pick for the play?" asks Isabella.

"I trust the group." I don't have a problem trusting everyone else, even though I'm not sure about Isabella.

I run to Sanya's house. Once I arrive, I manage to calm both her and Toby down.

Fatal error #1: I should have asked Sanya to watch Toby in *our* home, where he is more comfortable. Fatal error #2: I forgot to mention that his new favorite snack is cheese and crackers. Sometimes he can be flexible, but the changes in routine were more than he could handle. I should have known better. Now Sanya and Toby are both wrecks.

I tell Sanya I'll pay her extra once I get my allowance. Toby fixates on the crackers and cheese. I can barely get in the door without him demanding the food.

"Toby, I know you are mad. I shouldn't have left you with Sanya." I place his snack on the table in front of him.

"I want cheese and crackers."

"You have them, *see*?" I pick up a cracker and hold it close to Toby's mouth. He takes a bite and nearly catches my fingers. "Hey…" I want to snap at him, but that will upset him more. I need him to be calm when Mom gets home.

I watch him eat three more bites—he shoves the whole cracker in his mouth and barely chews. "Is everything okay now? We don't need to worry Mom with the details. *Right*?"

"What details?"

I jump. Mom's home early.

She drops her briefcase on the floor. Toby rushes toward her and lays his head against her chest. It looks awkward. Mom knows something is up.

"Okay, let me have it. What happened?"

I bite my nails and try to think of a cover story. Nothing comes. "Mom, I wanted to tell you yesterday. The school canceled drama. You know how much I want the class." I pause to catch my breath. "We decided to make it an extracurricular activity. Today was our first after-school session. I asked Sanya to watch Toby. I didn't think it would be a big deal."

I've been looking at the floor, but I glance up to see Mom's face. Her cheeks are flushed. That means I'm in trouble.

"This is not acceptable, Miranda. You know that your brother comes first. You promised me you'd make him your priority. Has something changed?"

"No," I whisper.

But something has changed. I want time for me. For my friends.

Why don't I say this out loud?

Toby takes his head off Mom's chest and moves back to the kitchen table to finish his cheese and crackers. They must be cold now, because he spits the food onto his plate.

"You can wait until drama is offered *during* school time." Mom washes her hands and begins to cut a steak into long strips to stir-fry. "We won't be getting Tobias a sitter. End of story. Understood?"

"I don't understand—it's not fair! It's not my fault drama was canceled. It's not my fault my brother can't take care of himself!"

"Don't you talk like that." Mom misses the meat and nicks her finger. She wipes it with a paper towel. My head becomes an angry fog. Toby doesn't have a clue what's happening. He's fine while I've lost drama before it even starts.

I refuse to join Mom and Toby for dinner. I'm too upset. I go to my room and call Amber. "Hey, it's Randi. How's the unpacking?"

"Nearly done! Did you hear what play we are doing?"

With all the stress around Toby, I forgot about choosing the play. "No." My shoulders slump. I nearly drop the phone. "What did the group decide?"

Her voice rises slightly as she answers. "Mr. Dean decided we should do a play by Shakespeare."

"What? I don't remember any Shakespeare being on the list." Just like that, I'm pumped again. There has to be a way I can make this work. "So," I say. "Which play is it?"

I have seen a few Shakespearean plays with Mom and Toby at Granville Island. The free performances showcase teen actors. Maybe that's what got me thinking about a future career in acting. I forget that I'm pissed at Mom and think about the shows we've seen. How we usually take a picnic and sit on the grass so Toby doesn't have to be too close to that many people. How my brother always laughs at the characters that bring slapstick comedy to the show.

"*A Midsummer Night's Dream.*" Amber's voice cuts through my thoughts.

"When Mr. Dean heard complaints that it wasn't a modern play, he said that this one is kind of like an urban fantasy. Somehow I doubt it will be relevant today, but it does have fairies and magic."

"Well, did you already pick parts?" I ask.

"No, we're supposed to look over the play and choose three parts we might like. Then each of us will give an audition based on our first choice."

My clothes feel too warm.

What part would I want? How can I be ready to audition by Friday?

What will I do with Toby?

"By the way," asks Amber. "Where did you go?"

"Ah…family emergency. Nothing serious, but I had no choice. Maybe tomorrow we can talk about the play?"

"Sounds good," says Amber. "I gotta go. My mom's been ragging on me about homework."

I don't care what my mom says. I'm going to take this class. I'm going to make this work. Now I have to decide if I'm brave enough to try out for a main part.

Mom hollers up the stairs, "I'm off to yoga. You need to watch your brother." I hear the front door slam shut.

When I return to the living room, I head for the bookshelf. I have to step around Toby, who is on the floor, building with his Lego set. I find Dad's old copy of *The Complete Works of William Shakespeare*. The book smells old and musty. The print looks like it was typed in a font size of four. I squint at the words and scan the list of characters. Then I remember this play from Granville Island. It was a lot of fun to watch the girl fights between the main characters, Helena and Hermia.

I randomly pick a section to work on and begin saying Helena's part out loud, one line at a time, until I can remember it. The words are strange. We don't talk this way. Will I even be able to remember these lines?

"Have you not set Lysander, as in scorn, to follow me, and praise my eyes and face?" Toby repeats as I finish a short monologue.

"Quit it, Toby. You'll get me confused." I don't know if I can do this. I don't want to make a fool of myself in front of Josh. And I still don't know what I'm going to do with Toby on Friday when we have the auditions.

It's after ten when I crawl into bed. I fall asleep thinking of Helena and how the boy she loves is interested in the other girl.

Chapter Six

At breakfast on Friday, I practice my lines between bites of toasted waffle. I don't dare say them out loud. Mom can't know I've decided to take Toby with me to the audition.

All through the walk to his school, I practice the phrases I've memorized. I still don't fully understand the text. I think Helena is upset that Demetrius doesn't love her and Lysander is making fun of her, but I'm not sure. I couldn't ask Mom to help me. So I'll have to wing it.

My concentration is off all day. My nerves are twisted up in my throat, and my mouth is so dry it's hard for me to swallow. I drink eight gallons of water, but it doesn't help.

After school I run to Toby's class. He's ready to go. I should have prepared him for the change in our routine, but I was afraid he'd have a bad day. Then I wouldn't be able to go to the audition at all.

"Toby, we are—"

"*Tobias*," he reminds me.

"Listen, you know the lines I've practiced, right?"

"*To call me goddess, nymph, divine and rare…*" he recites.

"Exactly!" How I wish those words were true. "I am going to a place where

I get to say those lines. Wouldn't it be fun to watch me?"

"You do your lines, and I will have cheese and crackers."

Since the disaster with Sanya, I've learned to be better prepared. I brought the supplies with me. I scoped out the backstage area of the theater earlier. There's a long table with a microwave plugged in at one end. I can easily warm his food, he can play on my iPad, and it will be fine.

"Yes, I have cheese and crackers for you. We will have them at *my* school. Remember coming to my school in the summer?"

"Your school and you do your lines, and I will have cheese and crackers."

"That's right," I say. I'm proud that I calm myself down so Toby doesn't pick up that I'm stressed.

As we cross the busy street two blocks from my school, Toby grabs my hand and pulls me in the opposite direction. We're in the middle of the road and the walk sign is counting down.

Fifteen, fourteen, thirteen…

Toby tugs harder in the opposite direction. I pull him toward the sidewalk so we can go to my school. Obviously, he misunderstood and thinks we are going home.

"*Toby*," I say, using a firm voice like Mom uses when she needs him to listen. "*It's not safe.* We must cross the street. We can't stay in the middle of the road." My pits drip with sweat.

Two, one… The light has now turned yellow. I pull Toby with all my strength, and he lurches forward. Because he's just about as tall as me and much heavier, I have to check my balance so I don't fall. Toby's hands flap as we finally step to the safety of the sidewalk. I look around. No one is paying attention to us.

"Toby, it's *okay*. You can play on my iPad. I have your cheese and crackers. We're almost at the school. You're hungry, aren't you?"

Even though I left before the bell at my school, with the walk back I'm late. I don't want to miss my audition. At homeroom this morning, I made sure Mr. Dean knew that I might be late and that I'd be bringing my brother. With all the students around, I couldn't explain my brother's needs.

Now I wish I had.

Toby walks with me, holding the edge of my coat. To feel safer, he'll stick to me like glue.

When we finally arrive at the theater, Isabella is onstage, giving her audition. I didn't know we'd be performing in front of the other students. I thought it would just be Mr. Dean! Sweat drips down the side of my body, and I'm aware that it has stained my light cotton shirt. I see Josh nod as I come down the aisle.

Everyone claps for Isabella, and Mr. Dean says hi to me.

"You and your brother can put your things on that side of the auditorium." He points in front of the stage, to the left.

That wasn't in my plans. I should say something. But with everyone's eyes on me, I'm not sure what to do, so I follow Mr. Dean's instructions. I deliberately sit farther back with Toby, away from the other students.

"I want cheese and crackers." Both the food and my iPad are at the front of the stage. Why didn't I take my things out first?

"You'll have to wait a few minutes," I whisper. My voice is tighter than normal. I wish I was allowed to concentrate on one thing. Getting the role.

Now I can't give Toby what he needs, and Isabella just auditioned for the same part as me!

After Mr. Dean gives feedback to Isabella, which sounds like high praise, he calls on me to come to the stage. Why didn't I practice both parts? Then I could have switched my audition to Hermia. Now I'll be compared to Isabella.

Toby is flapping so hard I'm sure he'll take off and fly around the auditorium. Everyone's eyes are on my brother as I take center stage.

I can see Toby rocking in his seat, and I can hear him muttering about dead people. Mr. Dean's eyes are on his notes, while everyone else in the theater is watching Toby.

Josh is the only one looking at me. He winks.

"And begin," says Mr. Dean.

I stumble forward to be closer to the edge of the stage. The lighting heats my body. A drop of sweat trickles down my face. When I blink, it feels like tiny fingernails cutting my eyes. I forgot my drops this morning.

What's my first line?

I hear myself stammer out a few words. They aren't the ones I meant to say. I have to start over. Why can't I ever come first in our family? People are still focused on Toby, who is flapping and mumbling.

I spew out my lines like I'm giving Toby and my mom hell for everything that's gone wrong today. The words flow, but I can hear the edge in my voice. I've said them too fast.

Now I've blown the audition for sure.

I hear laughter and realize Toby is repeating the lines I just said. I jump off the stage and grab our things. As I pass Toby's aisle, I reach over to his seat and yank him up hard. He winces but allows me to drag him up the aisle toward the door. I don't wait to hear Mr. Dean's remarks.

I know what he's going to say.

Chapter Seven

Tears stream down my cheeks as Toby and I bolt from the school. He's still flapping and making odd sounds. Mom will flip if she sees us like this. That will be the end of my acting career for sure. Maybe the end of my life!

To calm Toby, I detour to the park near our house. We sit under Toby's favorite tree. He loves this spot because the blue jays inhabit the branches and watch him. As they cackle, he cackles back. Soon his hands rest quietly by his side, and my breathing slows down.

I know I've blown the audition. Not only did my portrayal suck, but I didn't wait to hear Mr. Dean's comments. Not very professional. Josh won't want to pay attention to someone who *claims* to like acting. There's no way Mr. Dean will give me a second chance to audition.

Now Toby is happily chirping beside me, even without his cheese and crackers. I guess he's glad to be in familiar surroundings. At least we'll be able to go home now. But first I have to convince him not to say anything about the audition. "Toby, we have to go. It's time to see Mom. She'll be home from work. You can tell her all about your *school* day, right?" There is still an edge in my voice.

"I fed Hopscotch. He is fast. He runs around the class."

"That's a great thing to tell Mom. She'll love hearing about Hopscotch." I cross my fingers that he'll forget to mention anything about the audition.

When we get home, Mom is already cooking dinner, and when she asks where we were, Toby tells her about the park and the blue jays and Hopscotch. He doesn't say a word about what happened after school.

I help clean up after dinner and then go to my room. Dad's book of Shakespeare is sitting on my bed. I recite my lines with ease now that the audition is over. Why couldn't I have delivered my lines like this?

My cell vibrates. It's a text from Amber.

R U OK? she asks.

My thumbs move quickly over the keypad. Not gr8t.

I was worried, she writes back.

Sorry I bailed. Brother not good.

Want to come over?

Tx but 2 much homework. Once I start my homework, I realize I didn't even ask how her audition went.

At breakfast, Mom says, "Remember our fundraiser shift at the mall is at eleven."

I completely forgot today is Saturday and that I promised Mom I'd join her and Toby to collect donations for autism. I don't have the energy, but I agree to go as I figure it will keep my mind off my botched audition.

We've been at the fundraiser table for half an hour when I spot Josh in the grocery lineup. He's buying gum and a carton of chocolate milk.

"Hey, Randi." He smiles as he approaches our table. "Is your brother okay? He seemed upset yesterday at the audition."

Mom fires a glare my way. *Crap!* I'm in for it now.

"Thanks…for checking." It's like the wind has been knocked out of my lungs. "He's…okay. But I blew my audition worrying about him." I look at the sign behind me, hoping Josh will figure out that my brother has special needs and cut me a break.

"Cool that you look out for him. I think you were brilliant yesterday. You played Helena just right. Irritated and jealous. You'll get the part."

Josh isn't harassing me about my brother and he thinks I'm a good actress? "But what about Isabella?"

Our table shakes as Mom taps her foot—a sure sign she's angry.

"She was all right. But the emotion wasn't raw, like your portrayal. I bet you get it."

"What did Mr. Dean say?" As soon as I ask, I wish I hadn't.

"He was fine. He didn't say anything about your performance. He just called up the next person. I hope I land the role of Demetrius. I'd really like to play him, but I'd be happy with Lysander too. Wouldn't it be cool if we both get the parts we want? I'm Josh," he says to my mom.

"Oh yeah. Sorry I didn't introduce you. Josh tried out for the play."

Mom nods but doesn't say anything, which means I'm in a load of trouble.

"Well, see you at school on Monday," I say. He smiles and drops a toonie into the collection box. Toby says, "Thank you," like Mom taught him. I watch Josh leave the store. I wish I could join him so I don't have to face my mom's wrath.

Chapter Eight

"Miranda Woods, do you want to tell me what that was all about?" asks Mom as soon as Josh is out of hearing range.

"I…I didn't want you to get mad."

"I get *mad* when you deliberately go against my wishes."

A lady with thin gray hair interrupts our discussion to tell us about her son. "My boy, Stuart, was a good kid.

But in those days, people didn't understand autism. I didn't have a word for what he had until he was in his forties. Good that you know what your boy's troubles are. It will make it easier to get the help you need." She plops a ten-dollar bill in the jar. "You make sure you have breaks. Take advantage of those people who give respite. You'll need it!"

She walks with a limp as she leaves the store. Mom looks sad.

"She's right, you know." I choose my words carefully. "We need a break sometimes. What is respite anyway?"

"First of all, your brother is *not* a burden. He has needs that are different from other kids, but he is not a *problem*. *He's* the one we need to support."

Just like that, Toby is the center of our family once again.

"*Whatever*," I snarl.

"No, not *whatever*. I want you to remember your responsibilities. We all have jobs in this family, and you seem to forget that." Mom shakes her head at me.

"Toby doesn't have a job."

"Would you like to walk in his shoes?"

A family stops at the table, and the father asks Mom questions. He holds his daughter's hand. The table is busy for the next half hour, so we don't have a chance to talk. Mom looks exhausted. We focus on Toby playing games on my iPad for the last ten minutes of our volunteer shift. But once we're in the car, Mom lets me have it again.

I am to go straight home after school. She will call me on my cell phone to check in. Every day!

Do I argue with her? No.

Do I say anything in my defense? No.

Monday at breakfast, Mom reiterates the importance of understanding that my brother's needs come first. Like I haven't known that since Toby was born.

At school, Amber rushes up to meet me just before the bell rings.

"I missed you on the weekend. Do you think the parts are posted yet? Should we head to the drama room and see?"

My heart says no, but my feet start walking toward the drama room. Amber follows. There's a commotion around the door to the auditorium. On the bulletin board on the back of the door is a long green sheet with all of the roles listed.

Isabella is in front of us. "This must be a typo! Mr. Dean praised my audition. I'm going to speak with him right now!"

"But it's homeroom first block," says Amber. "You'll have to wait."

Isabella pushes Amber and storms down the hall. I wonder what's got her so riled up.

Josh slides in beside me. "Wow. There they are. The words are written. The parts are posted. Break a leg? Or break a heart?" He winks at me as he moves up to see the chart. "Yes! I got Demetrius. Oh, 'tis a good day!"

Amber and I laugh at his impersonation of Shakespearean speech. He grins widely and swooshes his arm like he is bowing before me.

"Check out your names and parts, fair maidens."

I shuffle toward the list. I look for Josh's name first. Yup, he got Demetrius, all right. Then I look at Amber's name. She will play one of the fairies attending Titania, Queen of the Fairies.

"Congrats, Amber. You got the part you want!" I stall some more by looking at Isabella's name and the character she got. I see now why she left in a huff. She was counting on playing Helena.

Instead, she has the role of Hermia. So who got Helena?

I scan the left side of the list until my eyes find Helena's name. I slowly read across the page. Beside the character of Helena is the name Miranda Woods.

"You deserve it," says Josh. "You nailed the audition."

He pats my shoulder and nods. Amber and I hug. I feel disconnected from my body. Can this be true? Did I really get the part? The bell rings for homeroom, and I glide down the hall, not feeling the linoleum beneath my feet. Amber is speed-talking about how she'll wear her hair for the part of fairy attendant. People high-five Josh as he enters our classroom.

I can't quit now. How can I make this work with Toby? And how am I ever going to keep this a secret from Mom?

At the end of the day, I leave early to pick up Toby. At least he doesn't blow going across the street as we head back to my school. Since I figured I wouldn't be in drama anymore, I didn't bring his snack. As I set him up with my iPad, I see that Mom gave him cheese and crackers for recess. He hasn't eaten them. Lucky break!

After I heat his food, Toby settles in to eating his snack. I can't wait to get my lines. I'm still in shock that I got a major part. That I got any part. Mr. Dean has adapted the play to make it fit our grades. Although I thought we would start by learning our lines, Mr. Dean surprises me by saying that we won't work on the play today.

"First we have to get to know each other." He asks us to sit in a circle on the stage. The floor is dusty. I have to remember not to wear black leggings on the days we have drama. Josh is across from me. He smiles. I bite my lip and grin back at him.

"The first activity is to get comfortable with how a scene is created. Think about the last time you were in the rain."

We laugh. It rains so often in Vancouver, we don't have to think hard to remember that. Toby loves the rain. He can stand at the window for hours, watching the drops run down the glass. I look out to the seats in the auditorium. Toby seems relaxed and absorbed in my iPad, so I let myself focus on the task from Mr. Dean.

"Now use your hands to create the wind."

Josh puts his hands together like he's praying. Then he blows air into them. It sounds like wind. I try it. Other students blow air across their hands.

"Good," says Mr. Dean. "Now tap the floor to make rain." Too many students pound the floor, and it sounds like a stampede of wild bulls. "*Soft* rain!" he says. We try again, and this time tap lightly. "Now a bit harder, to show the rain falling faster…faster…faster. Now it's a storm, with gale winds!"

Josh makes awesome wind sounds, and a few other students copy him. I pound the floor. It feels great. Together we're making a storm. I feel connected.

"And stop! Excellent work. Now the next activity. This one is to get you moving around. To be aware of your body. You are going to be beans." We stand, and Mr. Dean directs us to give one another space. Holding our arms out, we make sure we aren't touching anyone else. Josh has moved in beside me and now has to shift a few steps back.

"Okay. I'm going to call out a type of bean, and you respond with certain actions. For broad bean, you have to stretch your arms sideways, make yourself as wide as you can. For string bean, you have to stand as tall as you can." Several students mimic the bean by standing on their toes and reaching for the ceiling. Isabella pouts.

"For the frozen bean, freeze like a statue. For the jelly bean, wiggle and wobble all over. For the French bean, say, 'Ooh la la,' and for the jumping bean—well, that one is obvious. Any questions?" No one says anything. I can feel the electric energy. We want to move. To release some tension. And to play.

"Ready?" says Mr. Dean. I smile at Josh and nod. I'm ready.

"Jelly bean."

We wobble and fall down and laugh.

"French bean." More laughter and *ooh la la*'s.

"Frozen bean." The room stops moving. Everyone is silent. Everyone, that is, except for my brother. He is shouting from the seats in front of the stage, "OOH LA LA! OOH LA LA! OOH LA LA!"

The silence is broken. Everyone cracks up. Toby says the lines again.

"Stop it, Toby. Play your game." My voice is loud. Isabella shoots a dirty look my way.

"Ooh la la," says Toby in a loud whisper.

Mr. Dean shouts the next command to distract us, and we are back in the game. Toby returns to playing with my iPad. Mr. Dean covers all the beans two or three times, but he never goes back to the French bean. Toby has spoiled our fun.

Just before Mr. Dean sets us up for our next activity, my cell vibrates in my shirt pocket. It's Mom. I move behind the shabby curtain on the right of the stage so I'm out of range of the noise.

"How is your day going, Randi?"

"Good, Mom. Toby is having his snack." That's true. "We're just hanging out." In a way, that's true too.

"Glad to hear it. I'll be leaving work at four, so I'll be home shortly. See you soon."

Thank goodness she thought to let me know. And thank goodness we don't have a land-line. If we did, she would know we're not at home. Now I'll have to tell Mr. Dean that I have to leave early.

As I turn back to the group, Mr. Dean has already explained the next exercise. Josh fills me in that we are to pretend to eat toffee. Amber acts as though the toffee is stuck in her braces. Students exaggerate chewing.

I feel disconnected from the group after my mom's check-in. Instead of joining the activity, I watch my class-mates. Many are focused on Toby. They look at him to see what he's doing. They copy his motions. He pretends to chew and makes gross sounds.

Who is taking drama, my brother or me?

Chapter Nine

Toby and I are going to drama three times a week. I've managed to keep it a secret for two weeks now. During our last rehearsal we learned about expressing a range of feelings. Mr. Dean says it is important to know how to show the levels of emotion, like frustration, anger and rage. We also have to know how to get into a feeling and maintain control.

That way, we'll be able to project our lines and remember our actions.

When we were halfway through rehearsal, I got to practice controlling emotions. Mom called to check in, and I kept my cool and acted like Toby and I were at home. So far she hasn't caught on to my deception. But how long can I keep this up?

On the days we don't have drama I practice my lines after school. I'm getting better, but some words are tough to remember. My brother seems to know them better than I do. I only hope today I can say them without looking at my sheet.

Now Toby sits in front of the stage while he plays with my iPad. Mr. Dean gives us more activities that allow us to practice using our voices. He says I have to breathe from deep within, and that will help me to find my voice. Josh projects his voice so you can hear him at

the back of the theater. I still find it hard
to speak up.

It doesn't help that every so often
the group breaks into laughter as Toby
loudly exclaims from the sidelines,
"The zombies ate my brains."

I know they aren't laughing at him,
like other kids in school, but it still
stings. What makes it worse is that
I forget what I am supposed to say or
do every time Toby draws the group's
attention toward him. Twice I almost fall
off the stage because of Toby's timing.
Twice I bump into Isabella. She sneers
both times.

Mr. Dean allows us to work on our
character portrayals. He shares more
about the play but wants us to figure out
our characters on our own. Toby laughs
when Ward, who plays Snout, pretends to
be a wall. I see him put down my iPad
and study Ward. Ward makes his finger
and thumb resemble a hole in the wall.

That way, the characters Pyramus and Thisbe can speak through it. Toby copies him and pretends to talk through both sides of the hole. Then he laughs. Even though it's funny, I have to remind myself to focus on *my* character.

In the last half hour of our session, Mr. Dean says we'll rehearse a few scenes with our partners. But first he wheels out a large crate with a lid. Inside are costumes. "Take one item of clothing that seems to fit the part you are playing. It helps to wear something that allows you to jump into the character."

Everyone crowds around the crate. I stay back, waiting until it's less crowded. Amber shows me a light-purple scarf she's found. She wraps it around her waist. Then she holds up a weird-looking hat thing.

"What *is* that?" I ask.

"I think it's a bonnet," she says. "I'm going to sew some things this weekend that you and I can wear. The stuff in here is pretty lame."

I take the bonnet from her and put it on. Do these things ever get washed? Yuck! It smells like greasy hair and sweat.

Mr. Dean arranges us in pairs and groups of three to practice our lines. Amber slides in close to Sean. He plays Lysander. Since we've been doing the play for a few weeks now, I know that Helena likes Demetrius. I hope Josh and I will be together. But Mr. Dean places me with Isabella.

"Are you even ready?" She smirks. "I hope you know how to play your part, since you stole the character I want. And I don't get why *he's* here." She points at Toby. "Such a distraction!"

My body flips an *overheat* switch. Instead of using my own words, I open the book with my lines. I jump in where the words fit. "*Lo, is she one of this confederacy?*" My frustration fuels my speech. "*Have you conspired to bait me with this foul 'game'?*" I can't believe I'm breathing from my diaphragm. The words are loud and confident.

Mr. Dean passes by and says, "Good, Randi. Keep it up."

Isabella squeals, "*I scorn you not; it seems* you *scorn* me*!*"

As we spit the lines at each other, I realize our feelings match the play. I scan my sheet for the next line. I'm ready to deliver it when I see out of the corner of my eye that Toby is now on the stage, rifling through the costumes. "Toby, leave those alone."

"Focus on your lines," snarls Isabella.

Toby ignores me. He finds a donkey mask and places it over his head. He stumbles and lurches toward me. Students stop reading their parts to point and laugh. Toby can't see where he's going and nearly falls off the stage.

"Toby, get that off your head now!" I yell. Everyone is watching him.

"*You scorn me*," he says. The group laughs harder.

"Toby, take it off!" I shout.

He tries to pull off the mask, but it covers his whole head. He's doing it all wrong, pulling it away from his face instead of up and over his head. He pulls at it again. He stomps his feet and makes a loud wailing sound. He can't free himself. His arms flail. He shakes his head back and forth, as if trying to loosen the mask's grip.

"Get it off! Get it off!" he screams.

He spins out of control.

As I move to help him, Isabella steps in my way. "Will someone tell that crazy kid to *shut up*?"

"*You* shut up." My hands squeeze into fists.

"Get it off! Get it off!" Toby shouts again as he stumbles around.

I'm worried he can't breathe. I shift my attention from Isabella and move toward Toby. But Josh gets there first. He tugs on the mask, and Toby yells. He puts one hand on Toby's shoulder and with the other hand pulls off the mask. Toby punches the air with his arms. One arm connects with Josh's jaw. Hard.

Josh stumbles backward and falls to the floor.

I don't know whether to help Josh or Toby.

Toby's forehead and cheeks are red. He's sweating. His eyes are wide. "Get it off!" he shouts again. His arms are still wild. The whole group is staring.

Josh stands and brushes off his pants. A red mark appears on his chin where Toby hit him.

Before I can say or do anything, Isabella jumps in front of me. This time she's so close I can smell garlic on her breath from lunch. Her words bite.

"So *this* is how it's going to be? We can't even practice our lines. You better figure out what's more important"—spit lands on my face—"being an actress or spending your life *babysitting*. I don't think you're good at either."

I push her shoulder, and she falls down.

Mr. Dean moves toward my brother.

I yell at Toby, which only makes him flap harder. He's still screeching, "Get it off! Get it off!"

The group surrounds us.

Then Toby does something he's never done before. He pushes past the students and slides off the stage. Then he bolts from the theater.

Chapter Ten

My ankle twists as I jump from the stage to follow Toby. Josh runs down the stairs and takes off after my brother. I limp down the aisle. I hear Mr. Dean dismiss the class as Josh and Toby exit the theater. They are out of sight by the time I reach the door and leave the school.

Now I am in pursuit of both my brother and Josh. I don't know which way they have gone, but I assume Toby will retrace our steps. I head toward his school. The sun burns the top of my head, and my brain feels like it's on fire.

What a mess!

I glance at my watch as I approach the school doors. It's already four thirty. Mom will be on her way home.

I don't care that my foot hurts. I run down the hall to Toby's classroom. I'm sure he'll be there waiting. But the door is locked. The lights are off.

Oh no! Where would he go if he can't get into his class?

"Toby! Where are you? Toby, it's me…Randi." I call frantically. I run down each hall in the school. I check both bathrooms. He's not hiding in the stalls. He used to do that when he was younger. But he's not there. He's nowhere in the school.

I head to the back of the building. The playground is empty.

Panic explodes inside me. Where would he go?

I head to the park and his favorite tree. He's not there. My cell rings. It's Mom. Do I dare answer it? What if she's home and knows we're gone? What if she knows Toby is missing?

"Yes, Mom?" I say, but my voice is barely above a whisper.

"Do you have a cold?" she asks.

Mom wouldn't start with that question if she knew something was up with Toby.

"No," I answer. "Are you on your way home?"

"I have to stay late. That's why I'm calling. There's lasagna in the freezer. Thaw it out and have it for dinner. Are you sure you aren't coming down with something? You sound wheezy."

"No, I'm good. I'm really good," I say, catching my breath. "I have to go. Toby's hungry." I know that will be true. After his outbursts, he's always famished.

I close my cell phone and run as hard as I can. My ankle burns, but I have to find Toby. At least I have more time, with Mom staying late at work.

About a block from my house, I see Toby and Josh. My shoulders slump with relief, and I book it toward them.

"I am *so* glad you're here. Toby, you scared the hell out of me."

"Hell…you scared me." He places his head against my shoulder. I'm shorter than Mom, so he's actually leaning against my chin. I put my hands on his shoulders and pull him away so I can make eye contact.

"You are never to run away. Do you understand me? Never!"

"He was just scared," says Josh.

"I don't care if he was scared. He can't run away. What if you hadn't gone after him? He could have hurt himself."

"I get that you're upset," says Josh, "but yelling isn't going to help anyone right now."

Toby slides up beside Josh. "Crackers and cheese," he says softly.

Josh smiles. "I could use some crackers and cheese myself, buddy. You made me run fast chasing you."

"*You* ran fast?" I say, letting some of my anger turn to relief.

"You're limping. Are you okay?"

I put all my weight on my good leg.

It's hard to stay mad when Josh is being super sweet to my brother. I can't imagine what would have happened if he hadn't been with Toby. I decide to go easy on him.

"I'll make you both crackers and cheese."

Once in the house, the first thing I do is check Toby for any signs of injury. When he freaks out, he often hurts himself. He may scratch his face and arms, or he may hit himself in the head. This time, though, he's okay.

I grab ice from the kitchen before I start cooking or making snacks. Instead of taking it from me, Josh lets me place the ice against his chin. He puts his hand over mine. Electricity bounces through my body. I lower my eyes. The intensity in his eyes is exciting and scary at the same time. I break the moment by talking. "How come you're so good with my brother? Did you know he has autism?"

"Yeah. That day at the store. I saw the sign. But I sort of knew before." He takes the ice away from his face. As I sit beside him, he motions for me to put my injured leg up on the chair. I didn't realize how much it hurt until now. He places the ice on my swollen ankle and continues. "My cousin comes to my house every summer for respite. Bruce was born with Down's syndrome.

He's twenty-five. It's fun to hang out with him. When I was young, though, I didn't understand that he couldn't do everything I could. Like swimming."

"What happened? Did he almost drown?"

"No. Copying me ended up being the *best* thing that could happen. He learned to swim. Now he competes in the Special Olympics. I don't treat him any differently. So he tries everything I try."

"What is respite anyway?" I ask.

"It just means that my aunt and uncle are taking a break. They recharge while Bruce visits us. Then they have more energy for him. Don't you guys get respite?"

"I wish!" I look at Toby. The stress of the last hour is gone from his face, like he's already forgotten what happened. He hasn't even asked for cheese and crackers. Then I remember. He needs to eat, and I need to get the lasagna in the oven.

Josh and Toby play *Plants vs. Zombies* while I get the food sorted out.

"My mom won't be home until late, so if you want, you can stay for supper."

"Radical," says Josh.

"Radical," says Toby.

I think it's radical too.

Chapter Eleven

After dinner, Toby relaxes with his Lego. Josh suggests we use the time to practice our lines, since we will probably rehearse them next week.

We start with the chase scene. My character, Helena, is running after Demetrius. I read, *"O, I am out of breath in this chase! The more I pray, the less my grace. Happy is Hermia, wherever she lies, for she has attractive eyes.*" I pause. "Helena likes Demetrius, but both boys like Hermia. So Helena is feeling left out in this scene?"

"But a little more than that," replies Josh. "Helena feels like Hermia is conspiring against her. She feels like they are all being cruel to her."

"Kind of like bullying?"

"Totally," says Josh. "Let's start the scene again and add movement. Not full-out running or anything, but let's make it real."

My ankle is still swollen, but I agree. We work on our routine. After three practice runs, Josh starts with a new line. "*O me! You juggler! You canker-blossom! You thief of love!*" His voice is an octave higher than normal.

Toby laughs and repeats the line. His voice is just as high as Josh's. I giggle. Josh is reading Hermia's lines, not his own. I find the line in my book so I can reply. *"Have you no modesty? No shame? Fie, fie! You counterfeit, you puppet, you!"*

Josh replies, *"Puppet,"* and his voice squeaks, he's made it go so high. *"Ay, that's how the game goes!"* He skips ahead a few lines. *"How low am I? Speak! How low am I? I am not so low that my nails can scratch your eyes."*

"Yikes," I reply.

"That's not your line," says Josh.

"I know. But those will be dangerous words with Isabella yelling them at me. She'd love to scratch out my eyes."

"True." Josh still pretends to be Hermia and teases me by coming at me with his hands raised. I push him down, and we roll onto the floor, play-fighting. He's on top of me, laughing. Then the room gets quiet. I feel sparks go off in my body.

Josh brings his face closer to mine just as Mom comes through the door.

Everything that happened after school comes racing back to me.

Josh jumps up and says hello to my mom. Toby rushes to show her his Lego creation. I sit on the floor stunned. Josh just about kissed me. My brain can't compute everything.

"Hello, Josh. Tobias, I hope you had a good day. And Randi, did you get supper for your brother?" Mom is giving me "the look."

"I did, Mom. And I was...helping Josh practice for *his* play."

"Have you no modesty? No shame?" Toby repeats in a high voice.

Josh laughs, then straightens up when he sees the look on Mom's face. "I guess I should be going." He shrugs his shoulders in my direction. I wave and feel my face get hot.

After Josh leaves, Mom heads into the kitchen to warm up a plate of lasagna.

"So tell me, what's really going on?"

"Nothing, Mom. I was helping Josh learn his lines."

"Okay," Mom continues. "I'm glad to see you found a way to stay connected with acting. Josh seems like a nice boy, and Tobias is obviously comfortable around him."

I jump on that. "Toby likes Josh, Mom. That's a good thing, right?"

"Yes," she replies as Toby comes to sit on her lap.

"Well, maybe that means he'd be okay staying with someone else who is like Josh. Someone who 'gets' Toby."

"What are you saying?"

"Couldn't we sign up for respite— to give us a break? Give you and me a chance to do things together?" I speak fast so Mom can't interrupt and so I don't lose my nerve. Mom is silent. She looks sad. She ruffles Toby's hair, and he smooths it back into place.

The silence in the room is heavy. This is a big thing. Mom hasn't said no, but she isn't talking. Does this mean she's thinking about it?

"Where should we go this summer for our holidays?" Mom asks Toby.

"Disneyland," he shouts.

"Mom? What about what I just said?"

"I heard you. We have been doing just fine on our own. I know it's a lot on your shoulders. Disneyland sounds like a great way to reward our family!"

"Disneyland! You really think waving a theme park at me will make things better? That a vacation will make up for everything?" As if Toby could handle the crowds at Disneyland. What is she thinking?

Toby begins to rock in Mom's arms. I usually stop whatever I'm saying or doing if Toby gets upset. Now I don't care. Let him blow. Let him blast off like he did earlier. Let Mom see what's really going on!

My feet can't leave the kitchen fast enough. I take the stairs two at a time and then feel a shot of fire course through my injured ankle. I stumble on the last step and nearly tumble back down to the living room. I think about going to Amber's house. I could march past Mom and out the door. But my ankle hurts, and I just want to relax. I slip into my pj's and crawl into bed. I know it's early, but I am exhausted. Besides, it's quiet now.

I don't want to think about Mom and her lame offering of Disneyland. I don't want to think about my pain-in-the neck brother.

Maybe I can fall asleep thinking about Josh and our almost kiss!

Chapter Twelve

So much for dreaming about Josh last night. My only dream was a nightmare. Mr. Dean gave my part to Isabella. She and Josh got lost in a long kiss during rehearsal. And Mom banned me from returning to school, saying I would be homeschooled by a tutor.

I have never been so happy to wake up from a dream!

Since I have been practicing my lines and working on breathing from my diaphragm, I feel way more confident in my acting skills. And even though yesterday was Richter-scale earthquake stress, Toby doesn't say anything. He's quiet during breakfast and doesn't even read the obituaries out loud. Mom's biggest concern seems to be the kiss Josh and I almost shared.

I can't wait to get away from the table. I have so much to tell Amber about last night. I do all my Saturday chores. Mom has to work today, even though it's the weekend. And tomorrow she has a support-group session with other parents of kids with autism. She says they sometimes invite other family members, but she's never invited me. Guess she appreciates the built-in babysitter.

So as long as Toby doesn't spill his guts about yesterday, we should make it through the weekend. He usually forgets about things after a few days if he gets his needs met. So I'm going to be super careful with him and keep everyone happy.

I call Amber and invite her to come over for the afternoon. I've already talked with her about Toby's autism, since it's pretty obvious he's different. She thinks I'm really good with him.

"Hey, Toby," Amber says when she arrives. He's playing on my iPad and doesn't look up or answer her.

I don't bother teaching him how to act when someone says hello. We are supposed to always work on that, but I'm all about keeping things calm today.

"Looks like Toby is okay now," says Amber. "That was scary yesterday. I was worried he would hurt himself. And poor Josh."

I put my finger to my lips and say, "Shhh," so we don't get Toby riled up again. "What did Mr. Dean say after we left the gym?"

"Oh, he was fine. He totally understood why Toby freaked out. Then he dismissed us early for the weekend."

"So you don't think he's mad at me? You don't think he'll disqualify me from acting?"

"No, I think everything is cool. I often see Mr. Dean talking with Toby. I think everything will be fine."

"As long as my mom doesn't find out that Toby freaked," I say.

We gather the ingredients to make M&M cookies. Amber is about to dump all the M&M's into the batter when I stop her. I have to reserve some of the batter for just red M&M's for Toby, since they don't come in pink.

Once the cookies are ready, I tell my brother to sit at the table. "Toby, you can have four cookies and a glass of milk." I fill a glass and place it in front of him.

"Amber and I will be in the living room if you need me."

I pass Amber the plate of cookies and pour us two glasses of ginger ale, then lead us to the living room. I plop down on the sofa and Amber grabs a spot on the floor.

"I get my braces off next month. I can't wait." She uses her fingernail to dig out cookie crumbs from between the metal and her teeth. "They are such a pain."

"What's the first thing you'll eat when they are off?" I ask.

"Ice cream, popcorn, apples. But the more important question is, What's the first thing I'm going to *do* once they are off?"

"And?"

"I'm going to kiss Sean!"

We giggle. I pull the throw pillow close to my chest and hug it. I wish Josh and I could continue what we started yesterday.

Maybe he'll want to rehearse with me tomorrow.

After we have our snack, Amber opens her knapsack. She pulls out a long skirt made of clear material. It's soft like silk.

"This is for you to wear in the play," she says. "You can wear a skirt or pants under it."

I pull the sheer yellow skirt over my leggings. It swooshes as I move. "I love it! Thanks. I wish I had your talent. Any chance you can help me with my home-ec class? I'm going to fail if I don't figure out how to finish the pencil case I'm working on. Putting in the zipper is impossible."

Amber agrees to help me. We spend the afternoon talking about Sean and Josh.

Sunday, Josh comes over while Mom is at the support group. She figures she'll be gone for a few hours. So Josh and I have lots of time to rehearse.

After we've been saying lines for about thirty minutes, with Toby copying most of them, my brother announces that he wants to go to the pet store. Since Hopscotch became a part of his class, he's been obsessed with animals. Josh walks with us to the pet store. All the way there, Toby pulls on my arm to make me walk faster. He focuses on rabbit food and repeats the line from the play where the character Bottom is magically turned into a donkey. "*I could munch some good dry oats!*"

I'm embarrassed, but Josh laughs with Toby. I think Josh senses my frustration as we enter the pet store.

"I bet you don't get to shop for yourself very often," he says as Toby runs to the cage with rabbits.

"I'm okay as long as Toby is happy."

"But you deserve to be happy too. Your brother is fine. Look at him."

I look at Toby, who's squatting in front of the rabbit cage to be closer to them. A brown rabbit with long whiskers has poked his nose through the bars. Toby is laughing and scrunching up his own nose.

"Why don't you take fifteen minutes and go window-shopping? I'll keep an eye on Toby."

"Seriously?"

"Yeah, go! Enjoy yourself. We'll be right here." Josh smiles at me, then bends down to Toby's level. A sign says you can pet the rabbits, so Josh guides Toby's hand toward the closest rabbit. Toby squeals with delight.

"Thanks," I say as I touch Josh's shoulder. He takes my hand and squeezes it gently. For a moment, it looks like he might kiss my hand. I pull it away slowly and smile.

Two doors down from the pet store is a clothing store with all the popular name brands. I'm still grinning from Josh's touch and his generosity with my brother. As I hold a pair of faded jeans up to my body to see if they fit, I hear Isabella's voice.

"You were allowed out without your brother chained to your ankle?"

"Give it a rest, Isabella." I check my breathing. I won't give in to the anger boiling up in my chest.

"You know, it would be better if you babysat your brother on *weekends* so you can come alone to rehearsals on *school* days. He's such a pain."

"Actually," I say, "I watch out for Toby every day. He's not a pain, and in case you haven't noticed, until Friday he was pretty quiet during rehearsals. Maybe if you hadn't pushed my buttons, everything would have been fine."

"I pushed your buttons? Get over yourself!"

I take a deep breath. I can see myself in the mirror behind a rack of jackets. I look calm and confident. I choose to ignore Isabella.

A woman with severe hair and mean eyes puts her arm on Isabella's shoulder. "No clothes! Remember, you have to lose ten pounds before I will buy you any new outfits. And why would you be in *this* store? You know these are trash clothes."

Isabella doesn't look my way. Her shoulders drop, and she looks at the floor. Her mom steers her out of the store, Isabella's feet barely touching the ground she is moving so fast. The scariest part of this whole scene is that Isabella already looks skinny to me. Where does her mom figure an extra ten pounds are hiding?

While I'm proud of myself for standing up to Isabella, I feel bad for her now. Her mom is nasty.

Josh, Toby and I buy ice cream for our walk home. This has been one of the best weekends of my life. Toby has been fun to be around. The little break Josh gave me felt like a vacation, even though Isabella interrupted my time. And so far Toby hasn't said a word to Mom about Friday's mask freak-out.

If only we can keep it that way!

Chapter Thirteen

It's Monday morning and back to school. We only have breakfast to get through. Toby doesn't have the obituaries page open in the newspaper. Instead, he's mumbling about Hopscotch and how he wants a pet. Mom half pays attention as she makes his lunch. I wish she'd listen. After seeing him in the pet store, I'm sure Toby would be happy with a pet.

I'm relieved we made it through the weekend without the Friday incident coming up. Only twenty minutes to go, and I'll be home free.

Then my world crashes.

Mom finishes making Toby's lunch. She goes to the chair by the side door where Toby *always* leaves his knapsack. It's not on the chair. Mom looks frantic as she moves to the front door.

When Toby tore out of the theater on Friday, he left without his knapsack. When I rushed after him, I didn't grab it.

Toby not having his knapsack is catastrophic!

I hold my breath while I wait for Mom to decide how to handle this. I want to suggest that we not tell Toby. I can carry his lunch to his class. But that would mean letting Mom know I am aware that the knapsack isn't here. Better to play dumb. Cross my fingers. Hope for the best.

Yeah. Right.

"Tobias." Mom leans toward him. He drops his spoon into the cereal bowl. It's wet as he fishes it out. He ignores Mom and grabs a clean spoon from the container on the table.

I stand up. Maybe if I move quickly, I can be out of the kitchen and out of the firing range. But it's too late.

"Tobias, I am speaking to you." Mom is using her firm voice. Toby starts to flap his hands. He hates it when Mom uses that tone. He knows he's in trouble.

"Put the spoon in the dishwasher," he says, then gets up and places *both* spoons in the dishwasher. I slide my plate in while the door of the dishwasher is still open and brush around the back of my brother so I can head into the hallway.

"Don't you move a muscle, Miranda. I have a feeling this concerns you too. Tobias, where is your knapsack?"

All hell breaks loose.

Toby rocks and chants, "Get it off me, get it off me."

"What? Get what off you? There isn't anything on you." Mom puts her hand on Toby's shoulder, and he shrugs it off. He has his head bowed. He rocks on the balls of his feet and bites his lip. Hard. Small red drops trickle down his chin.

Mom moves to the counter and grabs some paper towel to dab at Toby's lip. "Shhh. It's okay," she says to him. Then she turns to me. "What is this all about? I'll ask again. Where is your brother's knapsack?"

"Need my knapsack to go to school. Hopscotch has a birthday. Need a knapsack. To take carrots. For the birthday. Need my knapsack."

"You hate your stupid knapsack! I'll carry the dumb carrots. JUST RELAX!" I yell the last part, and Toby covers his ears.

"Now look what you've done. You've upset your brother. I have a meeting at work. I *must* be on time. This is ridiculous. You need to sort—"

"For once we agree on something, Mom. This *is* ridiculous. I didn't do anything to upset my brother. I'm tired of missing out on things that I want to do. I'm tired of you always taking his side."

"What other side is there?"

"Oh my god! Do you even hear yourself?"

Mom doesn't say anything. Toby stops rocking. She looks at me and shakes her head. "It isn't supposed to be this hard. You are the mature one. You know what your brother needs, and yet you're focused on yourself."

"Maybe neither of us really knows what Toby needs. Maybe his needs are changing. Maybe you're too focused on your work to notice what's really going on. Maybe if you spent more time…"

Mom steps within inches of my face. Her mouth is a thin line, and I can tell she's holding back from screaming at me. "Whatever happened on Friday, it obviously included you *not* paying attention to your brother. Get yourself to school, Miranda. I will drive your brother today. And I will be phoning your school to find out what's going on."

"I—"

"End of discussion."

"But..." I fight the tears and bite my lip. I need to think fast. I can't lose drama. I can't lose this opportunity.

"You are grounded for the week. I'll make arrangements with my boss so I can be here to enforce this. Do you understand?"

"I understand that this is totally unfair. You don't know anything." I stomp out of the kitchen and slam the front door. I run to the sidewalk. I cough as the crisp air hits my throat.

My glasses fog up. I brush the moisture from my eyes.

I didn't grab a lunch or my purse. I didn't put in my contacts. I have no cell phone or cash.

Now I wish the magic in the play was real—then maybe I could make everything better!

Chapter Fourteen

I cry most of the way to school. Why didn't I go back into the house to get my stuff?

I remember that Amber is going to help me with my pencil case for home ec. It will be good to talk with her. She'll understand what I'm going through.

I head to the drama room first so I can grab Toby's knapsack. Mr. Dean and an older student are building a stand for our set. I don't want anyone to see me looking like this, so I turn to leave. Mr. Dean spots me and waves the boy away.

"Randi, please wait." He walks down the stairs of the stage and motions for me to sit beside him in the front row of seats.

I'm glad I am sitting beside Mr. Dean instead of standing in front of him. It's easier to keep my emotions in check. I'll have to remember to use that in my acting. Right—no acting this year. I better 'fess up now so Mr. Dean has time to replace me.

"Randi, I have to tell you what a fine actress you're becoming. Your timing is excellent, and you portray real emotion. Even your voice has improved—I sat halfway back in the auditorium the other day and I could hear you easily."

"That makes what I have to say even harder, Mr. Dean." I shuffle in my seat and look at the stage. My mom and Toby could have been sitting here in a few weeks, watching me. But not now. Then I say something unexpected. "What would happen if I, say, missed a few rehearsals?"

"Well, one rehearsal, we'd be okay. Two would be challenging, because we're beginning to work on the fine points of your character, and looking at blocking and other things. Three, well, because you have a main role and interact with so many other actors, I'd sadly have to look for a replacement."

My body slumps deeper into the chair. I fight back a new batch of tears. "I...I may not be able..."

"Listen, Randi, if this has to do with your brother and what occurred on Friday, please know he's always welcome."

"You've been great, letting him hang out with us. Amber said you often talk with Toby?"

"He loves the theater as much as you do. He has a good sense of which lines to repeat at certain times. Almost as if he understands the meaning of each line."

"Toby uses echolalia to learn language, which means he repeats what he hears and reads."

"Maybe it's helping him to have so much input from something he enjoys," suggests Mr. Dean.

I wonder if drama class is the reason Toby has stopped reading the obituaries in the morning. Maybe he's got something he likes better.

"Yeah. But none of that matters now. I'm grounded. I guess I blew it." I take a deep breath and hold it—the oxygen feels good. "I've been lying to my mom for weeks. I was supposed to take Toby home after school and not participate in drama until it's offered during school hours."

"Ah." Mr. Dean nods. He doesn't say anything. I guess he's thinking about my replacement. Isabella will get her wish to play Helena after all. And now I won't find out what blocking is all about.

"I'm really sorry, Mr. Dean. I should have been honest and found another way to work things out. I love acting, and I love this play."

"That's apparent, Randi. Your struggles with Toby, and maybe the recent events with your mom, have given you real feelings to tap into as an actress. The best actors are usually the ones who understand emotions. Those who have lived through hard times can get deep into their roles. I see the potential in you."

"I better go. I have a home-ec project due today. If I blow that, there's no telling what my mom will do next."

"Do me a favor. Tell your mom what all of this means to you. But don't focus on only *your* needs. I have a feeling your brother needs this time as much as you. Okay?"

"Sure. Thanks for everything." I stand and shake Mr. Dean's hand. I've never done that with a teacher before. I guess this feels more like a career than school.

I don't recall walking down the hall, but now I am at the home-ec room, where Amber is waiting.

"There you are. I texted you a hundred times. What's up? Are you okay?" She slides in beside me. She puts her arms around me, and I start to cry again.

"I'm grounded. Mom will be home after school every day, and I have to be home too. Mr. Dean knows I'm not going to make drama, so I'm sure he'll give my part to Isabella."

"No! That can't be! Let me talk to your mom. I'll help you."

"Thanks, but I have to face this on my own. I lied to my mom. Maybe if I had been honest and handled things differently, she would have given me a chance. But it's over."

Amber helps me finish the sewing project and supports me on the walk back to homeroom.

At lunch I sit with Josh. I try to explain why I won't be in drama anymore. "It's embarrassing to say this, but I'm grounded. I've been lying. My mom didn't know I've been going to drama, so I guess I deserve it."

Josh doesn't say anything for a few moments. "Listen, I've seen how much you care about your brother. Why don't we talk with your mom and see if I can watch him today after school, so you can still go to drama?"

I take his hand. "That is really sweet. You are…" I can't think of the right words, so I kiss his cheek. "I'll talk to my mom. I know it won't change things, but I'll do it. Maybe…no, I don't want to get my hopes up."

I tell Josh I'll call him later and let him know what's happening. I walk to the park beside our school for the last bit of the lunch hour. I want to be alone with my thoughts so I can figure out how to talk to Mom. I pull a photo out of my wallet of Mom and me at Kits Beach. I tell her what I feel and what I need. It goes very well. Everything that I say makes sense, and Mom doesn't interrupt me once. Not once.

Now if only I could say all of this for real.

Chapter Fifteen

I take my time walking home after school. On the way, I practice my lines. But not the ones I'll say to Hermia or Demetrius. The ones I'll say to Mom.

She and Toby are already home when I arrive. Toby rushes to me and says, "Hopscotch loves carrots. Hopscotch loves birthdays."

"I bet," I say. "Sounds like you had a good day." And I mean it.

"We need to talk," says Mom.

"I know." I pull out a chair at the kitchen table. Mom and Toby do the same. I grab an apple from the fruit bowl in the middle of the table. Toby grabs one too.

"I'm sorry," Mom and I apologize at the same time.

"You go first," says Mom.

She isn't blasting me with anger, so I take a deep breath and plunge in. "I shouldn't have lied. I didn't tell you about Friday because I would have had to tell you that I was taking Toby with me to drama classes. I let things get out of control. I was so happy after I got the part I wanted—a *leading role*, Mom. I kept lying so I could perform. I've never wanted something this much."

"What upsets me most," says Mom, "is that you put Toby's safety at risk."

She reaches across the table and rolls her fingers through my brother's hair. "But I have my own apology to make. On Sunday at the support group, they were discussing respite and how valuable it is for siblings to get a break too. Maybe I wasn't keen on it before because I needed to know that I could do this on my own."

I'm about to say something when Mom adds, "That *we* could do it on *our* own."

Toby crunches his apple. He laughs. I guess it's a private joke. Then he rubs his chin and says, "*I must to the barber's. I am marvelous hairy about the face.*"

"What's that about?" asks Mom.

"That's Bottom's line in the play."

"So you've been doing Shakespeare, hey?"

"Yeah. *A Midsummer Night's Dream.*"

"How did Tobias ever learn those lines?" she asks.

Toby jumps up and does his "wall" routine. Mom laughs. I can't help myself. I crack up too. "Mr. Dean, Josh, Amber—they've all been really cool with Toby. *He* likes drama, Mom." I look at Toby and say, "French bean."

"Ooh la la," he says.

Mom laughs again.

"See, it's not just about me. Toby enjoys it too. He really likes coming to drama class. Yes, I lied, and that was wrong. But we've been practicing the play for weeks now." I take another breath, and my words come again, just like I practiced. "Could Toby come with me? Can I finish the play? I promise you can ground me for the rest of the school year. It's just that there are a lot of people—"

Mom jumps in. "Counting on you?"

I nod.

Mom continues, "Well, there is too much going on at work right now for me to be out of the office. I'm needed there. Like you're needed in the play. You are still grounded on the days you don't have drama, and you absolutely must look after Toby on rehearsal days."

In that moment I realize something. "You know, Mom, I don't need to look *after* Toby anymore. He's getting older. But I will still look *out* for him. I've been doing that for years, and I'm pretty good at it!"

She smiles and ruffles my hair. "Let's hope we can make this work."

"Fingers crossed," I say.

"Fingers crossed," repeats Toby.

Chapter Sixteen

Josh comes over Tuesday after school so we can rehearse.

"So what is blocking?" I ask.

"It's where we need to stand when we're on the stage. Mr. Dean gives us directions on how to move. When we do the chase scene, he wants us to run up and down over the stand he created, so it looks like we're moving through the forest."

Josh takes me through all the actions he and I will do when we have scenes together. He even remembers some of the blocking Mr. Dean showed Isabella for when she and I are onstage.

Wednesday at rehearsal, Isabella and I are working on our scene. We jump to the part where she and I are going to fight. Mr. Dean wants to go over the blocking, since I was away Monday.

Isabella doesn't look at me as we do our lines. I haven't seen her since the mall. Twice she stumbles over her words. She bites her lip and looks worried. I say my lines. Isabella doesn't respond. Then she taps her head and calls, "Line." That's what you do when you can't remember what your character is supposed to say. Before Mr. Dean can find the line in our book, Toby calls out, "*Puppet? Ay, that's how the game goes?*"

Isabella repeats the line Toby gives her and continues her speech. The room has gone quiet. She looks at my brother and whispers, "Thanks."

We are on our way to the theater. It's opening night, and I feel fireworks going off inside my body. Josh and I get to hold hands in the wedding scene. For the last few weeks, Isabella has been much calmer. We aren't friends, but at least she isn't stressing me out.

Since the day Toby fed Isabella her line, he's become our official prompter. He will stand on the side of the stage, behind the curtains, and quietly call out lines for any actors who forget them. Turns out he knows all of the play by heart. He even has a special costume to wear. Amber made it for him. And there's no mask. As Mom wheels the suv into the school parking lot, I turn to my brother. "So are you excited, Toby—

I mean, Tobias?" I figure it's time to show him a little respect.

He surprises me by replying, "Toby...*Sir* Toby."

Mom laughs. "That's the name of a character in another Shakespeare play, isn't it? I wonder where he got that."

"Probably from Mr. Dean," I say.

As we cross the parking lot, I'm thinking of the promise Mom made, that if Josh can watch Toby on Sunday, she and I will go to the support group together. She says the group helped her to see how much stress I might be under as one of Toby's main supports. And she has decided I'm no longer grounded. So as we enter the building, I'm feeling stoked about my future.

Then I overhear a family talking about Toby.

"Hey, guys, did you hear there's an autistic kid who helps out backstage?"

I turn to the family and say, "His name is *Toby*, and he's not an *autistic kid.* He is a boy *affected with autism.*" Now I finally get why Mom always says these words to me. I look at the family staring at Toby and add, "But he's way more than that. He cares about rabbits, and he can recite the whole play. He's my brother."

They look embarrassed and head into the auditorium. Toby and I let Mom find her seat while we hurry backstage. As well as being a fairy attendant, Amber has made most of the costumes. I grab mine and get dressed. Mr. Dean looks more nervous than the cast does. When it comes to my turn, I enter stage left. I see a blur of people. I suddenly feel overwhelmed. Everything that has happened since the first day of school comes rushing back to me. I begin to sweat, and even a deep breath doesn't help me remember Helena's first speech. I look at my brother and quietly say, "Line."

Right on cue, he tells me what to say. I have no problem remembering the rest of my lines. Isabella and I are able to pour a lot of emotion into our fight scene. She's a good actress, and I have fun playing my character opposite hers. Josh and I have a blast running up and down the stand in our chase scene. I almost tumble at one point but manage to keep my balance. He winks at me as he says his lines.

Once the play is over, we take turns coming out to the front of the stage to bow. Josh and I come out together, and Josh kisses my cheek, then takes my hand for our bow. He even stands aside so I have a chance to bow again on my own. Mom stands and claps. I'm too far away to know for sure, but I think she has tears running down her cheeks. For *me*!

Then I take Toby's hand and lead him out for the group bow. Mom claps again. The audience gives us all a standing ovation.

After the play, Mom takes us for hot chocolate. "I loved watching you in school performances when you were young. But you have matured into a strong actor. You have matured in so many ways." She ruffles my hair.

I smile. It's nice to have her support.

When I get home, I will put the program from the play in my special-memories book. It lists Tobias Woods as *Prompter*, and Miranda Woods as *Helena*. My first major role. As an actress, that is! It doesn't list *Sister*, but now I can be just as proud of that role.

Acknowledgments

Thanks to my editor, Melanie, for taking such great care with my novel. I want to honor all the past students with whom I've worked, who inspired me to write this story. I learned so much from your courage and strength. Thank you, Sean, for eagerly participating in Young Shakespeareans all those years ago, skillfully portraying memorable characters!

Cristy Watson is a teacher who enjoys reading and writing. The closest thing to theater that she does is host open-mic readings at her local coffee shop. Cristy lives in Surrey, British Columbia.

orca currents

9781554694341 9.95 PB
9781554698882 16.95 LIB
9781554698899 PDF • 9781554698905 EPUB

In most ways, Poe is like the other kids in his school. He thinks about girls and tries to avoid teachers. He hangs out at the coffee shop with his best friend after school. He has a loving father who helps him with his homework. But Poe has a secret, and almost every day some small act threatens to expose him.

Titles in the Series

orca currents